# No Metal No Magic

## Element 89

## Actinium, Presented By Acamus From The Magical Elements of the Periodic Table Book Series

Acamus

| 89 | 227 |
|---|---|
| **Ac** | |
| actinium | |

Actinium

*By Sybrina Durant with Illustrations by Pranavva et al.*

# No Metal No Magic

## Element 89

## Actinium, Presented By Acamus

## From The Magical Elements of the Periodic Table Book Series

Story copyright 2025

Soft Cover Print KDP—9798242090873

BISAC Codes:

JNF051070  JUVENILE NONFICTION / Science & Nature / Chemistry

JNF016000 JUVENILE NONFICTION / Curiosities & Wonders

JNF051080 JUVENILE NONFICTION / Science & Nature / Earth Sciences / General

Soft Cover Print ISBN 13 - ISBN: 978-1-942740-57-5

# Acamus The Actinide Knight Presents Actinium

This Element 89 book features the periodic table element, Actinium. It is presented by Acamus, a member of the Actinide Knights. Each knight has a magical sword or other medieval weapon tipped with an element that gives them unique powers. Their powers are based on the properties of its periodic table element.

Acamus is just one of the 118 elementals who will present all of the Magical Elements of the Periodic Table to readers who are curious about the wonders of the world.

Acamus introduces Actinium in his book.

The Actinide Knights and their other techno-magical friends are the perfect group to introduce you to the elements in the Periodic Table. Hopefully, this Magical Elements of the Periodic Table book will spark an interest in the magical and real world properties of all the elements known today. You may be surprised at how prominently they feature in our every day lives.

Each page in this book contains terms that might not be completely familiar to the reader. Refer to the definitions in the back of the book to get a clear understanding of each meaning.

There is also a fun elemental themed Periodic Table at the back of the book. It features 118 elements presented by fanciful characters like unicorns, dragons, wizards, knights and goblins.. They want you to remember that if there's no metal...there's no magic or technology.

Remember, "No metal – No Magic. . .and No Technology".

It's Techo-Magical.

Note: Sybrina Publishing websites are Sybrina.com and MagicalPTElements.com. Follow sybrinapublishing on Instagram, Magical Elements of the Periodic Table on Facebook, @sybrinad on Pinterest, Sybrina_SPT on Twitter; and Sybrina Durant on LinkedIn.

# Acamus The Knight With The Actinium Sword

Symbol: Ac  Atomic Number: 89  Atomic Mass: 227

Actinium resides in Group 03, Period 7 on the Periodic Table.

Magical elementals from the Magical Elements of the Periodic Table books present all of the elements of the periodic table in fantastical and real life terms.

In the books, each elemental character has magical powers based on the properties of the elements that come from the land, air and water. They are the perfect group to introduce you to metals, metalloids, non-metals, halogens, noble gases and much more.

Unicorns, dragons, alchemists, knights, and goblins will show you how people of this world always have and always will depend upon the elements that our earth provides for all of our needs.

Use this Periodic Table as you would any other to spark an interest in the magical and real world properties of elements known today. You may be surprised at how prominently they feature in our every day lives.

**Magical Elements of The Periodic Table**

No Metal

Actinium To Zirconium

No Magic

### 89  227
### Ac
### actinium

Acamus

Radioactive Medicine

It's Techno-Magical

The full periodic table of elements with magical character illustrations.

SUPER HEAVY METALS—RADIOACTIVE

RARE EARTH LANTHANIDE METALS

ACTINIDE METALS

## LEGEND

| Color | Category |
| --- | --- |
| | Alkali Metals |
| | Alkali Earth Metals |
| | Transition Metals |
| | Post-Transition (or Other Metals) |
| | Metalloids |
| | Non-Metals |
| | Halogens |
| | Noble Gases |
| | Rare Earth Lanthanide Metals |
| | Actinide Metals |
| | Super Heavy—Radioactive |

Alloys are created when 2 or more metals are combined. Compounds are created when 2 or more non-metals are combined.

**EXAMPLE OF A COMPOUND**

Quincy

Quick Lime =

Used for Concrete

Both Carbon and Oxygen are reactive nonmetals.

**White Wing**

Used for jewelry, dental amalgams plus connectors, and switch and relay contacts for electronics.

**EXAMPLE OF AN ALLOY**

Includes 58.5 % gold, 22% copper, 8% zinc, 7% nickel, 4.5% silver and possibly other elements.

White Gold =

Sybrina.com

~ 2 ~

# Actinium is a Actinide Metal

Actinium was discovered in 1899 by Andrew Debierne in Paris. It was first extracted from a rock called pitchblende, which contains uranium.

Actinium is a soft, shiny white metal that gives off radiation. It quickly reacts with oxygen and moisture in the air, creating a white layer called actinium oxide that stops it from reacting further.

Actinium is very radioactive, which makes it shine in the dark with a pale blue light. This light comes from the air around it, which is charged by the tiny particles it gives off.

Actinium is a good conductor of electricity and heat though its extreme radioactivity limits practical electrical uses.

Actinium is paramagnetic. Paramagnetism happens because some electrons in an atom or ion are unpaired. These unpaired electrons make a magnetic force that lines up with an outside magnetic field.

Actinium is an Actinide Metal.  It is the first element of the Actinide Series.

## LEGEND

| |
|---|
| Alkali Metals |
| Alkali Earth Metals |
| Transition Metals |
| Post-Transition (or Other Metals) |
| Metalloids |
| Non-Metals |
| Halogens |
| Noble Gases |
| Rare Earth Lanthanide Metals |
| Actinide Metals |
| Super Heavy—Radioactive |

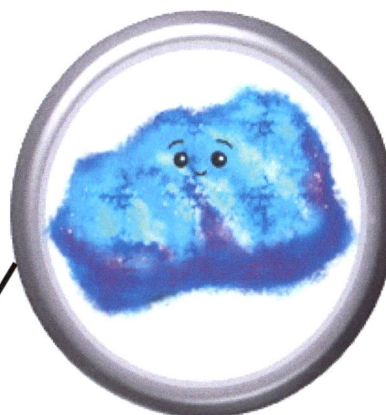

**Actinium Element**

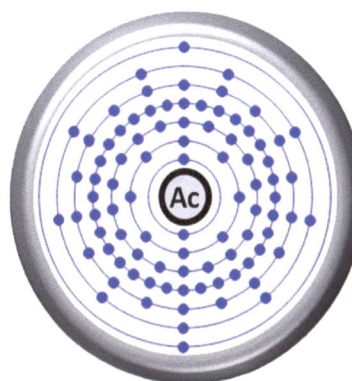

**Atomic Structure**

**Actinide Metals**—Any of a series of chemically similar metallic elements with atomic numbers ranging from 89 (actinium) to 103 (lawrencium). All of these elements are radioactive, and two of the elements, uranium and plutonium, are used to generate nuclear energy. The lanthanides and actinides are sometimes called the inner transition metals, referring to their properties and position on the table. They are actinium, thorium, protactinium, uranium, neptunium, plutonium, americium, curium, berkelium, californium, einsteinium, fermium, mendelevium, nobelium, and lawrencium.

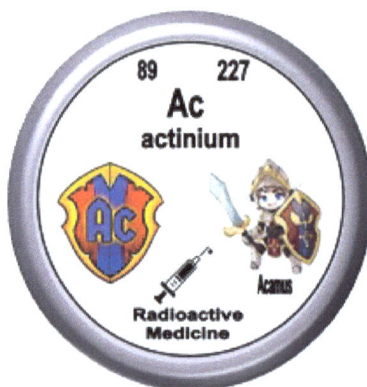

Actinium is a special element found on the periodic table. It has the symbol "Ac" and is number 89 in the list. It was discovered over a hundred years ago, in 1899, by a scientist named Friedrich Oskar Giesel, a chemist from France. He extracted it from leftover materials called pitchblende, which were left behind when Marie and Pierre Curie discovered radium. Actinium is part of a group called the actinides, which includes other elements like uranium and plutonium. These elements are important because they can be used in things like nuclear power and weapons.

Many scientists have studied actinium and learned a lot about it. One interesting fact is that actinium is a silvery-white metal and shines brightly. However, it is not something that people see often because it is rare and not found freely in nature. Instead, it is usually found in small amounts combined with other minerals.

One of the coolest things about actinium is that it is radioactive. This means that it gives off energy in the form of radiation as it breaks down over time. Being radioactive allows scientists to use actinium in medicine, especially in a type of treatment called targeted alpha therapy. This therapy can help treat certain cancers by targeting and destroying cancer cells while aiming to keep healthy cells safe.

Actinium exists in different forms called isotopes, which are versions of the same element but with different weights. The most common isotope of actinium is actinium-227. Researchers are particularly interested in this isotope because of its properties and potential uses.

Due to its radioactivity, working with actinium requires special care. Scientists use special equipment to keep themselves safe from radiation. Even though it can be dangerous, when used correctly, actinium can also have many benefits, especially in medicine.

Actinium is important in many fields. For instance, in the world of science and research, it has sparked interest due to its fascinating properties. At the same time, some industrial uses have emerged that utilize actinium's radioactive qualities.

There are ongoing studies trying to understand more about actinium and how it can be used safely and effectively. Scientists are exploring new ways to use it in treatments for diseases, which gives hope to many people looking for cures. Researchers are also looking into how actinium is formed in nature and its role in the Earth's geology.

To sum it up, actinium is a unique element with both valuable properties and potential dangers. Its shiny metallic appearance, combined with its radioactivity, adds to its allure. In medicine, science, and industry, actinium plays an essential role, providing tools and treatments that can help improve lives. As science continues to advance, we can expect to learn even more about actinium and how to use it in exciting new ways in the future!

# Uses For Actinium

There are very few current uses for Actinium but as the first element in the actinide series, it shares many chemical properties with the lanthanides, particularly lanthanum. Forward thinkers propose some interesting commercial potentials for its use in the future.

In the future, cars could be very different from the ones we drive today. They might use special materials called Actinium isotopes to help clean the air inside the car. These cars could have big, advanced air filters that keep the air safe and fresh. This way, passengers can enjoy a cleaner and healthier ride in their futuristic vehicles.

Actinium-powered energy sources could be used in places that are not connected to the main power grid. This means they can provide electricity in remote areas where people need power for things like lights, refrigerators, or phones. These energy sources are helpful for homes, farms, and other locations that want to be self-sufficient and independent from traditional electricity supplies.

# Uses For Actinium

(Continued)

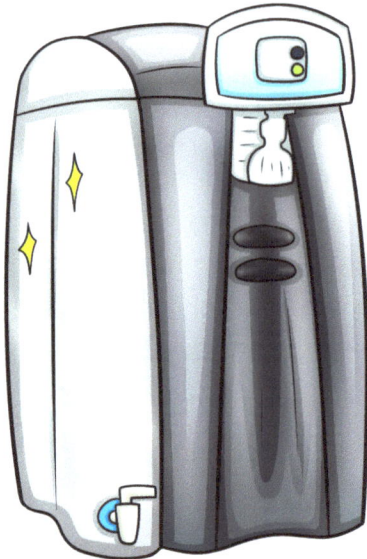

Actinium isotopes could be really helpful for cleaning dirty water in the future. These special forms of actinium might break down harmful substances and pollutants in water supplies, making them safe for people to drink. By using these isotopes, we might be able to improve water quality and protect our health. This is important because clean water is essential for a healthy life.

Actinium isotopes are special forms of the element actinium that can produce energy. These isotopes could be very useful in military robots that work on their own, allowing them to travel long distances without needing to stop for fuel. This means the robots could complete important missions without risking the safety of soldiers in dangerous areas.

# Uses For Actinium

(Continued)

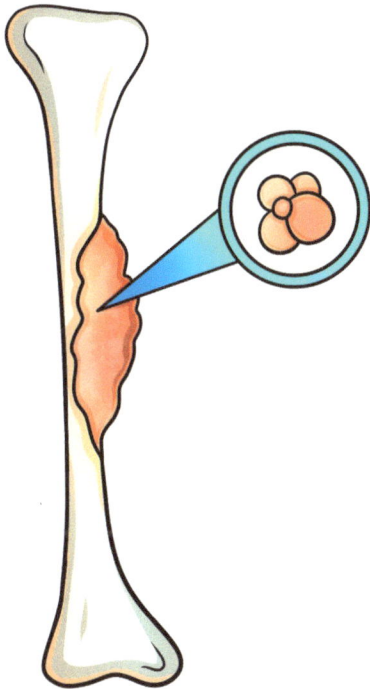

Actinium-225 is a special kind of material that doctors use to help treat certain types of cancer, especially in the bones. This material can find and attach to cancer cells, like how a key fits into a lock. By doing this, it helps to destroy the harmful cells while leaving healthy ones alone, improving the chances of recovery.

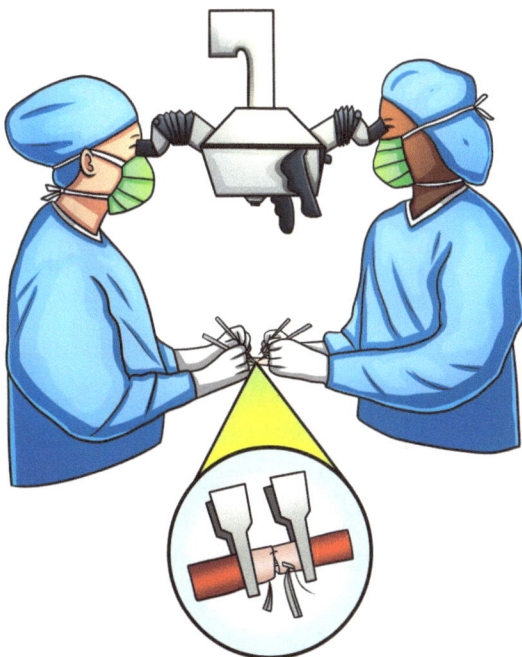

Scientists are currently studying actinium isotopes to find new ways to treat many illnesses using radiation. These isotopes might help doctors provide safer and less painful treatments for patients. The goal is to develop methods that deliver radiation directly to the area that needs help, while causing less damage to the surrounding healthy tissue. This research could change how we fight diseases.

# The Source of Actinium

Actinium is not found in large quantities in nature but some common ores are Uraninite and Thorite. It is primarily produced through the neutron bombardment of radium in nuclear reactors. The radium absorbs a neutron and undergoes beta decay to form actinium.

Actinium is a rare and somewhat mysterious element that belongs to a group known as the actinides in the periodic table. It has the symbol Ac and atomic number 89. Actinium is classified as a radioactive metal and is not something that you will find easily in nature. Because of its specific properties, it is primarily produced through artificial means, particularly in nuclear reactors and particle accelerators where high-energy processes can take place.

To understand how actinium is obtained, we first need to look at uranium and thorium, the two elements that are crucial to the process. Both uranium and thorium are naturally occurring radioactive elements found in small amounts in the Earth's crust. These elements go through a series of decay processes, slowly breaking down into different elements and isotopes over time. During this decay process, actinium is formed as a byproduct. This means that actinium does not exist on its own in significant quantities in nature, but rather, it is created as uranium and thorium decay.

Most commonly, actinium is produced by bombarding radium, which is another radioactive element, with neutrons in a nuclear reactor. This process is called neutron bombardment. When radium is hit with enough neutrons, it absorbs one and undergoes a change through a process known as beta decay. During beta decay, the radium transforms into actinium. This reaction occurs in highly controlled environments in nuclear facilities that can handle such radioactive materials safely.

After actinium is formed in these reactors, it is not immediately ready for use. The next step involves separating it from the other radioactive materials that were created during the decay process. This is where chemical separation and purification come into play. The nuclear reactor

# The Source of Actinium (continued)

produces a mixture of different isotopes, and actinium must be carefully extracted and purified. This extraction process is complex and requires a combination of chemical reactions, centrifugation techniques, and other methods to ensure that only actinium is isolated from the mixture.

The primary location where actinium is produced is within the confines of specialized nuclear reactors. These facilities are designed to handle high-energy reactions and radioactive materials safely. The processes that occur inside these reactors are highly controlled and monitored to prevent any radioactive leaks or accidents. The neutron bombardment of radium typically occurs in a section of the reactor known as the neutron flux area, where the concentration of neutrons is sufficient to enable these reactions to take place.

The extraction of actinium from the uranium ores involves additional steps. Uranium ores may be found in large mines, and these ores are processed in facilities known as mills. The milling process involves crushing the ore and using chemicals to separate uranium from the surrounding rock and other materials. This concentration of uranium allows for more efficient radiation treatments that lead to the production of actinium. Once the uranium has been concentrated, it undergoes further chemical processing, which allows for the extraction of actinium through various separation techniques.

One of the reasons why actinium is considered to be among the most expensive elements is the combination of its rarity in nature and the complexity involved in its production. The entire process of extracting actinium is costly, requiring sophisticated machinery and handling techniques to ensure safety due to the radioactive nature of the materials involved. Additionally, because of its rarity and the high level of technical expertise required to produce it, actinium is often used in specialized applications rather than for widespread use.

In conclusion, the production of actinium is a fascinating and intricate process that takes place under very specific conditions. This rare element is obtained primarily through the decay of uranium and thorium in nuclear reactors, particularly involving neutron bombardment of radium. The subsequent steps of chemical separation and purification occur in controlled environments, ensuring that the radioactive actinium is extracted from the mix safely. While actinium may be rare and expensive, its unique properties make it valuable for various applications in science and medicine, where its radioactive characteristics can be harnessed for beneficial research.

# Magical Elements of The Periodic Table

Magical elementals from the Magical Elements of the Periodic Table books present all of the elements of the periodic table in fantastical and real life terms.

In the books, each elemental character has magical powers based on the properties of the elements that come from the land, air and water. They are the perfect group to introduce you to metals, metalloids, non-metals, halogens, noble gases and much more.

Unicorns, dragons, alchemists, knights, and goblins will show you how people of this world always have and always will depend upon the elements that our earth provides for all of our needs.

Use this Periodic Table as you would any other to spark an interest in the magical and real world properties of all the elements known today. You may be surprised at how prominently they feature in our every day lives.

**No Metal**

**No Magic**

Actinium To Zirconium

Remember, "No Metal— No Magic." . . .And no technology.

| Group | # | Symbol | Mass | Element | Character | Real-world use |
|---|---|---|---|---|---|---|
| 1 | 1 | H | 1.008 | hydrogen | Hildy | Textile Manufacturing |
| 1 | 3 | Li | 6.94 | lithium | Lillian | Batteries |
| 1 | 11 | Na | 22.99 | sodium | Sorn | Salt |
| 1 | 19 | K | 39.10 | potassium | Pearl | Saline Drips |
| 1 | 37 | Rb | 85.47 | rubidium | Ruby | Night Vision Glasses |
| 2 | 4 | Be | 9.012 | beryllium | Berwyn | Musical Instrument |
| 2 | 12 | Mg | 24.31 | magnesium | Maggie | In Your Bones |
| 2 | 20 | Ca | 40.08 | calcium | Verly | Teeth |
| 2 | 38 | Sr | 87.62 | strontium | Strauna | Computer Screens |
| 3 | 21 | Sc | 44.96 | scandium | Sandra | Bicycles |
| 3 | 39 | Y | 88.91 | yttrium | Yago | Microwave |
| 4 | 22 | Ti | 47.87 | titanium | Tilly | Aerospace |
| 4 | 40 | Zr | 91.22 | zirconium | Zora | Chemical Pipelines |
| 5 | 23 | V | 50.94 | vanadium | Vana | Black Printer Ink |
| 5 | 41 | Nb | 92.91 | niobium | Konash | Mag Lev Trains |
| 6 | 24 | Cr | 52.00 | chromium | Crowmist | Stainless Steel |
| 6 | 42 | Mo | 95.95 | molybdenum | Maximo | Cutting Tools |
| 7 | 25 | Mn | 54.94 | manganese | Mangar | Earth Movers |
| 7 | 43 | Tc | 98 | technetium | Tephen | Radio Active Diagnosis |
| 8 | 26 | Fe | 55.85 | iron | Iown | Bicycle Chains |
| 8 | 44 | Ru | 101.1 | ruthenium | Ruth | Electrical Switches |
| 9 | 27 | Co | 58.93 | cobalt | Corliss | Magnets |
| 9 | 45 | Rh | 102.9 | rhodium | Rovana | Finish for Jewelry |
| 10 | 28 | Ni | 58.69 | nickel | Nix | Guitar Strings |
| 10 | 46 | Pd | 106.4 | palladium | Paedin | Concert Flute |
| 11 | 29 | Cu | 63.55 | copper | Cuprum | Money |
| 11 | 47 | Ag | 107.9 | silver | Silubra | Ventilator |
| 12 | 30 | Zn | 65.38 | zinc | Dr. Zinko | Suntan Lotion |
| 12 | 48 | Cd | 112.4 | cadmium | Cadmus | Power Tools |
| 13 | 5 | B | 10.81 | boron | Bordolas | Sports Equipment |
| 13 | 13 | Al | 26.98 | aluminium | Alumna | Airplanes |
| 13 | 31 | Ga | 69.72 | gallium | Gallant | LED Displays |
| 13 | 49 | In | 114.8 | indium | Iker | Liquid Crystal Display (LCD) |
| 14 | 6 | C | 12.01 | carbon | Cole | Charcoal |
| 14 | 14 | Si | 28.09 | silicon | Silonar | Glass |
| 14 | 32 | Ge | 72.63 | germanium | Gemlet | Camera Lense |
| 14 | 50 | Sn | 118.7 | tin | Tinam | Liquid Crystal Display |
| 15 | 7 | N | 14.01 | nitrogen | Nitra | Food Packaging |
| 15 | 15 | P | 30.97 | phosphorus | Phiova | Fertilizer |
| 15 | 33 | As | 74.92 | arsenic | Arkyn | Poison |
| 15 | 51 | Sb | 121.8 | antimony | Antz | Flame Resistant Fabric |
| 16 | 8 | O | 16.00 | oxygen | Ozzy | Air |
| 16 | 16 | S | 32.06 | sulfur | Xoe | Matches |
| 16 | 34 | Se | 78.97 | selenium | Selenice | Printers |
| 16 | 52 | Te | 127.6 | tellurium | Tellan | Vulcanized Rubber |
| 17 | 9 | F | 19.00 | fluorine | Fleure | Strong Bones and Teeth |
| 17 | 17 | Cl | 35.45 | chlorine | Krystia | Swimming Pools |
| 17 | 35 | Br | 79.90 | bromine | Brogach | Photography Film |
| 17 | 53 | I | 126.9 | iodine | Jody | Cloud Seeding |
| 18 | 2 | He | 4.003 | helium | Hetha | Balloons |
| 18 | 10 | Ne | 20.18 | neon | Jalan | Advertising Signs |
| 18 | 18 | Ar | 39.95 | argon | Arag | Light Bulbs |
| 18 | 36 | Kr | 83.80 | krypton | Krypto | Detect Leaks |
| 18 | 54 | Xe | 131.3 | xenon | Xena | Used To Catch Speeds |

## Period 6 (row 6) – Transition / Post-Transition Metals

| 72 178.5 Hf hafnium — Hallam — Nuclear Submarines | 73 180.9 Ta tantalum — Taaltra — Mobile Phones | 74 183.8 W tungsten (Wolfram) — Wolfie — 3D Printing Nozzles | 75 186.2 Re rhenium — Renkin — Rocket Engines | 76 190.2 Os osmium — Osm — For Lab Testing | 77 192.2 Ir iridium — Iridina — Weight Scale | 78 195.1 Pt platinum — Paedra — Pacemaker | 79 197.0 Au gold — Ghel — Pacemakers and Stents | 80 200.6 Hg mercury — Questa — Barometer | 81 204.4 Tl thallium — Thanein — Tattoo Ink | 82 207.2 Pb lead — Lauda — Batteries | 83 209.0 Bi bismuth — Bitsy — Fire Sprinklers | 84 209 Po polonium — Polgam — Anti-Static Brushes | 85 210 At astatine — Aszrad — Thyroid Cancer Treatment | 86 222 Rn radon — Ramarai — Earthquake Prediction |

## SUPER HEAVY METALS—RADIOACTIVE

| 104 267 Rf rutherfordium — Rolokz — Radioactive | 105 268 Db dubnium — Dubnic — Radioactive | 106 269 Sg seaborgium — Starx — Radioactive | 107 270 Bh bohrium — Boudak — Radioactive | 108 277 Hs hassium — Holga — Radioactive | 109 278 Mt meitnerium — Mohdort — Radioactive | 110 281 Ds darmstadtium — Dardank — Radioactive | 111 282 Rg roentgenium — Rogmort — Radioactive | 112 285 Cn copernicium — Clanvolt — Radioactive | 113 286 Nh nihonium — Nirtak — Radioactive | 114 289 Fl flerovium — Fleth — Radioactive | 115 290 Mc moscovium — Molit — Radioactive | 116 293 Lv livermorium — Ligee — Radioactive | 117 294 Ts tennessine — Tubnulk — Radioactive | 118 294 Og oganesson — Obyrit — Radioactive |

## RARE EARTH LANTHANIDE METALS

| 57 138.9 La lanthanum — Lannion — Telescope Lens | 58 140.1 Ce cerium — Cerella — Lighter Flint | 59 140.9 Pr praseodymium — Praethic — Welder Mask | 60 144.2 Nd neodymium — Nelushla — Electric Car Motors | 61 145 Pm promethium — Proctor — Night Light | 62 150.4 Sm samarium — Saranda — Electric Guitar Pickup | 63 152.0 Eu europium — Euell — Fluorescent Light | 64 157.3 Gd gadolinium — Galoa — MRI Diagnosis | 65 158.9 Tb terbium — Tarfin — Solid State Device | 66 162.5 Dy dysprosium — Dypsie — Sonar Sensors | 67 164.9 Ho holmium — Holmia — Eye Laser | 68 167.3 Er erbium — Erbie — Optical Communications | 69 168.9 Tm thulium — Thurwin — Eye Laser | 70 173.0 Yb ytterbium — Yitzy — Amplifies Fiber Optics | 71 175.0 Lu lutetium — Umii — Positron Emission Tomography (PET) |

## ACTINIDE METALS

| 89 227 Ac actinium — Acamas — Radioactive Medicine | 90 232.0 Th thorium — Thordis — Heat Resistant Paint | 91 231.0 Pa protactinium — Protik — Radioactive Waste | 92 238.0 U uranium — Uri — Used To Power Submarines | 93 237 Np neptunium — Neptisos — Nuclear Fuel | 94 244 Pu plutonium — Plodian — Power Satellites | 95 243 Am americium — Amerine — Smoke Detector | 96 247 Cm curium — Curran — Moon Rover | 97 247 Bk berkelium — Beremere — Scientific Research | 98 251 Cf californium — Calestian — Metal Detector | 99 252 Es einsteinium — Elizama — Nuclear Research | 100 257 Fm fermium — Farley — Scientific Research | 101 258 Md mendelevium — Menessant — Scientific Research | 102 259 No nobelium — Norhnum — Nuclear Research | 103 266 Lr lawrencium — Larelis — Radioactive Research |

## Alkali / Alkaline Earth Metals

Row 6:
- 55 132.9 Cs caesium — Caelkoth — Atomic Clocks
- 56 137.3 Ba barium — Barsena — Spark Plugs

Row 7:
- 87 223 Fr francium — Franchie — Cancer Treatment
- 88 226 Ra radium — Reele — Luminous Watches

57 thru 71 RARE EARTH LANTHANIDE METALS
89 thru 103 ACTINIDE METALS

# It's Techno-Magical

## EXAMPLE OF AN ALLOY

White Gold

White:
- 30 65.38 Zn zinc
- 47 107.9 Ag silver
- 79 197.0 Au gold (Gold)
- 28 58.69 Ni nickel
- 29 63.55 Cu copper

Includes 58.5% gold, 22% copper, 8% zinc, 7% nickel, 4.5% silver and possibly other elements.

White Wing

Used for jewelry, dental amalgams plus connectors, and switch and relay contacts for electronics.

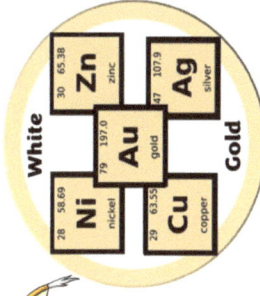

Alloys are created when 2 or more metals are combined. Compounds are created when 2 or more non-metals are combined.

## EXAMPLE OF A COMPOUND

- 20 40.08 Ca calcium — Verty
- +
- 8 16.00 O oxygen — Ozzy

Quick Lime = Teeth ... Air

Used for Concrete

Quincy

Both Carbon and Oxygen are reactive nonmetals.

Sybrina.com

## LEGEND

- Alkali Metals
- Alkali Earth Metals
- Transition Metals
- Post-Transition (or Other Metals)
- Metalloids
- Non-Metals
- Halogens
- Noble Gases
- Rare Earth Lanthanide Metals
- Actinide Metals
- Super Heavy—Radioactive

# All Of The Periodic Table Elements Listed Alphabetically

| | | | |
|---|---|---|---|
| ACTINIUM—*AC*—89 | EINSTEINIUM-*ES*-99 | MERCURY *(QUICK SILVER)* — *HG*—80 | RUTHENIUM—*RU*—44 |
| ALUMINUM—*AL*—13 | EUROPIUM—*EU*—63 | MOLYBDENUM—*MO*—42 | RUTHERFORDIUM—*RF*—104 |
| AMERICIUM—*AM*—95 | FERMIUM—*FM*—100 | MOSCOVIUM—*MC*—115 | SAMARIUM—*SM*—62 |
| ANTIMONY—*SB*—51 | FLEROVIUM—*FL*—114 | NEODYMIUM—*ND*—60 | SCANDIUM—*SC*—21 |
| ARGON—*AR*—18 | FLUORINE—*F*—9 | NEON (Jazzy)—*NE*—10 | SEABORGIUM—*SG*—106 |
| ARSENIC—*AS*—33 | FRANCIUM—*FR*—87 | NEPTUNIUM—*NP*—93 | SELENIUM—*SE*—34 |
| ASTATINE—*AT*—85 | GADOLINIUM—*GD*—64 | NICKEL—*NI*—28 | SILICON—*SI*—14 |
| BARIUM—*BA*—56 | GALLIUM—*GA*—31 | NIHONIUM—*NH*—113 | SILVER—*AG*—47 |
| BERKELIUM—*BK*—97 | GERMANIUM—*GE*—32 | NIOBIUM—*NB*—41 | SODIUM—*NA*—11 |
| BERYLLIUM—*BE*—4 | GOLD—*AU*—79 | NITROGEN—*N*—7 | STRONTIUM—*SR*—38 |
| BISMUTH—*BI*—83 | HAFNIUM—*HF*—72 | NOBELIUM—*NO*—102 | SULFUR (Xanthous)—*S*—16 |
| BOHRIUM—*BH*—107 | HASSIUM—*HS*—108 | OGANESSON—*OG*—118 | TANTALUM—*TA*—73 |
| BORON—*B*—5 | HELIUM—*HE*—2 | OSMIUM—*OS*—76 | TECHNETIUM—*TC*—43 |
| BROMINE—*BR*—35 | HOLMIUM—*HO*—67 | OXYGEN—*O*—8 | TELLURIUM—*TE*—52 |
| CADMIUM—*CD*—48 | HYDROGEN—*H*—1 | PALLADIUM—*PD*—46 | TENNESSINE—*TS*—117 |
| CALCIUM (Vital)—*CA*—20 | INDIUM—*IN*—49 | PHOSPHORUS—*P*—15 | TERBIUM—*TB*—65 |
| CALIFORNIUM—*CF*—98 | IODINE *(JODIUM)*—*I*—53 | PLATINUM—*PT*—78 | THALLIUM—*TI*—81 |
| CARBON—*C*—6 | IRIDIUM—*IR*—77 | PLUTONIUM—*PU*—94 | THORIUM—*TH*—90 |
| CERIUM—*CE*—58 | IRON—*FE*—26 | POLONIUM—*PO*—84 | THULIUM—*TM*—69 |
| CESIUM—*CS*—55 | KRYPTON—*KR*—36 | POTASSIUM—*K*—19 | TIN—*SN*—50 |
| CHLORINE (Keen)—*CL*—17 | LANTHANUM—*LA*—57 | PRASEODYMIUM—*PR*—59 | TITANIUM—*TI*—22 |
| CHROMIUM—*CR*—24 | LAWRENCIUM—*LR*—103 | PROMETHIUM—*PM*—61 | TUNGSTEN—*W (WOLFRAM)* — 74 |
| COBALT—*CO*—27 | LEAD—*PB*—82 | PROTACTINIUM—*PA*—91 | |
| COPERNICIUM—*CN*—112 | LITHIUM—*LI*—3 | RADIUM—*RA*—88 | URANIUM—*U*—92 |
| COPPER—*CU*—29 | LIVERMORIUM—*LV*—116 | RADON—*RN*—86 | VANADIUM—*V*—23 |
| CURIUM—*CM*—96 | LUTETIUM (Unique)—*LU*—71 | RHENIUM—*RE*—75 | XENON—*XE*—54 |
| DARMSTADTIUM—*DS*—110 | MAGNESIUM—*MG*—12 | RHODIUM—*RH*—45 | YTTERBIUM—*YB*—70 |
| DUBNIUM—*DB*—105 | MANGANESE—*MN*—25 | ROENTGENIUM—*RG*—111 | YTTRIUM—*Y*—39 |
| DYSPROSIUM—*DY*—66 | MEITNERIUM—*MT*—109 | RUBIDIUM—*RB*—37 | ZINC—*ZN*—30 |
| ERBIUM—*ER*—68 | MENDELEVIUM—*MD*—101 | | ZIRCONIUM—*ZR*—40 |

# Acamus Presents
# Actinium

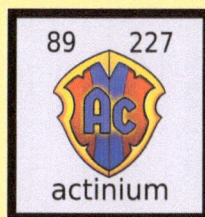

| 89 | 227 |
|----|-----|

**Ac**

actinium

## Did You Know?

*Actinium isotopes could be used in autonomous military robots for long-range missions.*

Magical Elements of The Periodic Table

- Actinium is the very first element in a group called the actinide series. The name "actinide" comes from the word "Actinium," its name and the reason for the group.

- Actinium was first discovered in 1899 by André-Louis Debierne, a French chemist. In 1902, a scientist named Friedrich Otto Giesel also independently discovered it. He named it emanium but that name did not stick since it had already been named by Debierne. Giesel can rightfully be credited with the first preparation of radiochemically pure actinium and with the identification of its atomic number 89.

- The word "actinium" comes from the Greek word "aktis," which means "sunbeam." This name was chosen because actinium is very radioactive, like how a bright sunbeam can be strong and powerful.

- Actinium-227 emits beta particles with so much energy that it lights up the surrounding environment with a blue glow. In a sense, due to its high levels of radioactivity, it is vibrating faster than the speed of light and the blue glow is like a sonic boom.

- The only naturally occurring isotope of actinium is 227Ac. Thirty-six radioisotopes of actinium have been identified, all with half-lives ranging from 69 nano seconds at the shortest (for 217Ac) to 21.77 years at the longest (227 Ac).

- Actinium-225, is being used in a new type of cancer treatment called targeted alpha therapy. As the atom decays, it releases particles that destroy the nearby cancer cells, but don't travel far enough to damage the rest of the body.

# Structure Of Elements In The Periodic Table

Periodic tables are laid out in rows and columns.

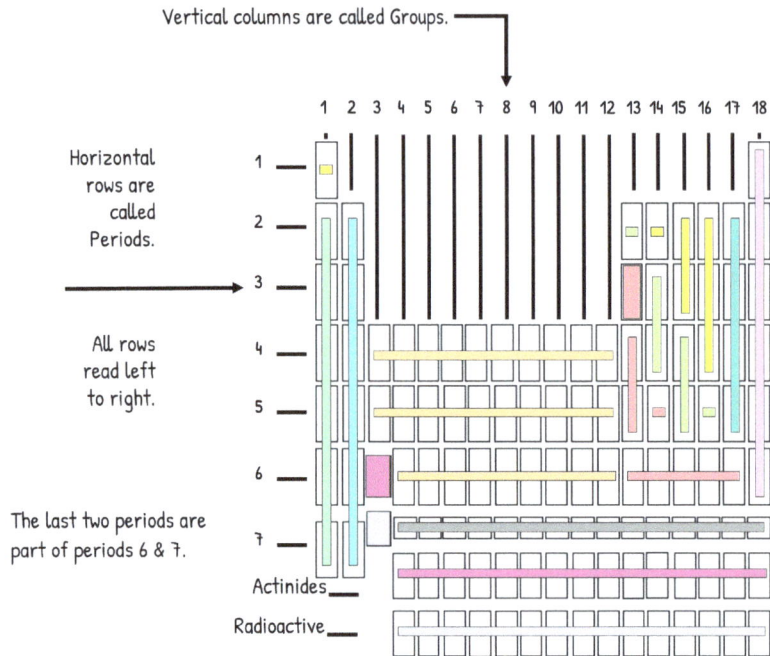

Vertical columns are called Groups.

Each element is placed in a specific location because of its atomic structure. Elements are arranged in Families.

Horizontal rows are called Periods.

All rows read left to right.

The last two periods are part of periods 6 & 7.

Actinides___
Radioactive___

The Families are:

- Alkali Metals
- Alkali Earth Metals
- Transition Metals
- Post-Transition (or Other Metals)
- Metalloids
- Non-Metals
- Halogens
- Noble Gases
- Rare Earth Lanthanide Metals
- Actinide Metals
- Heavy—Radioactive

The term 'Element' is used to describe atoms with specific characteristics.
Every element in the first column or Group has 1 electron in the outer orbital (shell).
Every element in the second column (group two) has two electrons in the element's outer orbital.
The number designation of each Group represents the number of electrons in the element's outer orbital—
except for Group 18, Period 1—Helium. It only has 2 electrons.
Those electrons, called Valence Electrons, are what chemically bond with other elements.

**Atomic Structure of Element:** The atomic structure of an element refers to the arrangement of protons and neutrons in the nucleus of the atom, and the electrons in the electron cloud around the nucleus. Group 1, Period 1—Hydrogen is the only element that has no neutrons.

Ac —Actinium
Group 3—Period 7

Atomic Number— Indicates the number of protons in the nucleus and the number of electrons in the atom (Element).

Atomic Mass (or Weight) — Equals the number of Protons + Neutrons in the Element's nucleus.

Atom (Element)

Electrons

89    227

## Ac
actinium

Element Symbol— The identifying letters for the element are an abbreviation of the element name. Only one or two letters are used for element symbols.  Ac= Actinium

Protons and Neutrons in Nucleus

**Atomic Structure**

Acamus

The Actinide Knight introducing this Element in the Magical Elements of the Periodic Table Book is Acamus the Knight With The Actinium Tipped Sword.

## Radioactive Medicine

Element Name—
The name of this element is Actinium. It is a Actinide.

An every day use for the element.
IE: Actinium is used for scientific research.

# Types of Elements On The Periodic Table

**Alkali Metals**—Some metals on the periodic table are soft and shiny. They are so soft that they can be cut with a knife! These metals are excited to give away electrons to elements in need, making them highly reactive. Ther electron transfer creates a compound known as a salt. Surprisingly, these metals are not found in nature alone; they must be extracted from other sources. Examples of these metals include lithium, sodium, potassium, rubidium, cesium, and francium.

**Alkali Earth Metals**—The elements in column 2 of the periodic table have 2 outer electrons in their shell. Ther makes them very active with nonmetals that need electrons to stay stable. When they react, they make something called a salt. They are often found in nature all by themselves, and they can even conduct electricity. The elements are beryllium, magnesium, calcium, strontium, barium, and radium.

**Post-Transition (or other Metals)**— Elements directly to the right of the transition metals. They are known as "poor metals: and are soft and brittle. These include aluminum, gallium, indium, tin, thallium, lead, bismuth, zinc, cadmium and mercury.

**Transition Metal**—The main metals are found in the middle and bottom rows of the periodic table. They look like metal, can conduct electricity, can bend and be shaped easily. The period 4 transition metals are scandium, titanium, vanadium, chromium, manganese, iron, cobalt, nickel, copper, and zinc. The period 5 transition metals are yttrium, zirconium, niobium, molybdenum, technetium, ruthenium, rhodium, palladium, silver, and cadmium. The period 6 transition metals are lanthanum, hafnium, tantalum, tungsten, rhenium, osmium, iridium, platinum, gold, and mercury. The period 7 transition metals are the naturally-occurring actinium, and the artificially produced elements rutherfordium, dubnium, seaborgium, bohrium, hassium, meitnerium, darmstadtium, and roentgenium.

**Metalloids**—The elements called metalloids are a mix of metals and nonmetals. They look like metals, but can't conduct electricity very well. They also break easily and act like nonmetals. These include boron, silicon, germanium, arsenic, antimony, tellurium, astatine, and polonium.

**Non-Metals**—These elements reside in columns 15-17, and can be gases, liquids, or solids. They don't conduct heat or electricity. The solids are brittle, and they have no metallic luster. They readily accept electrons from metals to form salts. These include nitrogen, oxygen, fluorine, chlorine, bromine, and iodine.

**Halogens**—Halogen chemicals are a special type of element. When they mix with metal, they become a kind of salt. Halogens are super reactive because they like to take an electron from metals. They can be found in column 17 of the element table. Some of them can be found in nature, but most are very dangerous and can hurt you if you touch them. They include fluorine, chlorine, bromine, iodine, and the radioactive elements astatine and tennessine.

**Noble Gases**—These elements reside in column 8. They are all odorless, colorless gases that are chemically very stable (inert). They don't generally form compounds by bonding with another element. These include helium, neon, argon, krypton, xenon, and radon.

**Lanthanide Rare Earth Minerals**—The Japanese call them "the seeds of technology." The US Department of Energy calls them "technology metals." These elements have atomic numbers 57-71. They are vital to industry. They can be added to metals to strengthen them to make alloys such as stainless steel, used to refine crude oil, and are crucial in producing technology—electronics, telecommunications, and metal devices to name a few. They are lanthanum, cerium, praseodymium, neodymium, promethium, samarium, europium, gadolinium, terbium, dysprosium, holmium, erbium, thulium,

**Actinide Metals**—Any of a series of chemically similar metallic elements with atomic numbers ranging from 89 (actinium) to 103 (lawrencium). All of these elements are radioactive, and two of the elements, uranium and plutonium, are used to generate nuclear energy. The lanthanides and actinides are sometimes called the inner transition metals, referring to their properties and position on the table. They are actinium, thorium, protactinium, uranium, neptunium, plutonium,

**Super Heavy—Radioactive**—Superheavy elements are those elements with a large number of protons in their nucleus. Elements with more than 92 protons are unstable; they decay to lighter nuclei with a characteristic half-life. They do not occur in large quantities (if at all) naturally on earth, and only exist briefly under highly controlled circumstances. They include lawrencium, rutherfordium, dubnium, seaborgium, bohrium, hassium, meitnerium, darmstadtium, roentgenium, copernicium, nihonium, flerovium, moscovium, livermorium, tennessine, and oganesson.

# Alpha-Emitting Transuranic Actinide Elements

These elements are all synthetic, meaning they don't occur naturally on Earth, and are produced through nuclear reactions. They are all radioactive and decay by emitting alpha particles, among other decay modes. Transuranic or transuranium elements can be classified as technogenic nuclides, meaning it has been produced and released into the environment due to human nuclear activity.

These four elements are the most abundant and the most extensively used of the man-made actinide series elements. They comprise a major radioactive waste disposal concern because they are long lived and have high radiotoxicity. Neptunium and Plutonium are the only transuranium elements that have been found in trace amounts in nature.

## Neptunium
2,100,000 Year Half Life

## Plutonium
24,000 Year Half Life

## Americium
458 Year Half Life

## Curium
17.6 Year Half Life

These seven elements are produced in such small amounts, mostly for research purposes; and most of the isotopes produced have such short half-lives, a few seconds or minutes, that they are an unlikely health concern.

  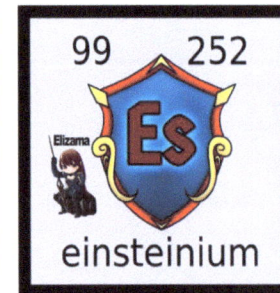 

## Berkelium
## Californium
## Einsteinium
## Fermium

## Mendelevium
## Nobelium
## Lawrencium

# Radioactive Isotopes

All elements have **isotopes**. There are Stable Isotopes and there are Radioactive Isotopes which are known to be Unstable. Every chemical element has one or more radioactive isotopes. More than 3,000 Radioactive Isotopes (radioisotopes) are known, of which only about 84 are seen in nature. We will discuss some of them here.

### Uranium =

3 Naturally Occurring Radioisotopes:

Uranium-238

Uranium-235

Uranium-234

Uranium has 24 man-made radioisotopes.

| 92 | 238.0 |
|---|---|
| **U** | |
| uranium | |

**Used To Power Submarines**

<u>U-238</u> Radioisotope Applications: The half-life of uranium-238 is approximately 4.468 billion years. It's alpha decay begins with thorium-234. This is the most common isotope of uranium found in nature. It can be used to generate plutonium-239, which itself can be used in a nuclear weapon or as a *nuclear-reactor* fuel supply.

### Radium =

4 Naturally Occurring Radioisotopes:

Radium-223

Radium-224

Radium-226

Radium-228

Radium has 34 man-made radioisotopes.

| 88 | 226 |
|---|---|
| **Ra** | |
| radium | |

**Raele**

**Luminous Watches**

<u>Radium-223</u> Radioisotope Applications: Has a half-life of 11.43 days and is part of the actinium decay series. It *targets bone tissue with alpha particles*, which have a short path length, reducing toxicity to adjacent healthy tissue. It can control painful bone metastases, delay complications like fractures, and improve survival.

### Francium =

2 Naturally Occurring Radioisotopes:

Francium-221

Francium-223

Francium has 32 man-made radioisotopes

| 87 | 223 |
|---|---|
| **Fr** | |
| francium | |

**Francine**

**Cancer Treatment**

<u>Francium-221</u> Radioisotope Applications: Francium-221 has a half-life of 4.8 minutes. When it undergoes alpha decay, it loses 2 protons and 2 neutrons, resulting in the formation of astatine-217. Due to its rarity and extreme instability, has no commercial applications and is primarily used for research purposes in fields like chemistry and atomic structure, including *spectroscopic investigations*.

### Polonium =

4 Naturally Occurring Radioisotope:

Polonium-210

Polonium has 41 man-made radioisotopes

| 84 | 209 |
|---|---|
| **Po** | |
| polonium | |

**Polgarn**

**Anti-Static Brushes**

<u>Polonium-210</u> Radioisotope Applications: Polonium-209 has the longest half-life of all polonium isotopes at 124 years. It is formed as a decay product of uranium-238. It is primarily used in industrial devices to eliminate static electricity, particularly in processes like *paper rolling*, plastic sheet manufacturing, and spinning synthetic fibers. It's also used in neutron sources and dust removal brushes for photographic films and camera lenses

Can you guess the most commonly used radioisotope?

| 43 | 98 |
|---|---|
| **Tc** | |

= ??????

(Answer can be found below.)

The above chart only shows a few of the radioactive isotopes formed by those elements. Other very important radioisotopes are:

**Iodine-131**: Used to diagnose and treat various diseases associated with the human thyroid. **Molybdenum-99 (Mo-99)**: Used as the 'parent' in a generator to produce technetium-99m. **Americium-241 (Am-241):** Used in smoke detectors, and other applications like thickness gauges and neutron sources.

The most commonly used radioisotope is *Technetium-99m (99mTc)*, employed in over half of all nuclear medicine procedures, used for imaging various organs and tissues. It is used in tens of millions of medical diagnostic procedures annually

# Definitions

**Atomic Structure of Element:** The atomic structure of an element refers to the arrangement of protons and neutrons in the nucleus of the atom, and the electrons in the electron cloud around the nucleus.

**Atomic Number:** An element's atomic number refers to the number of protons it has in its nucleus. In a neutral atom the number of protons always equals the number of electrons.

atomic number: 6

number of neutrons: 6

**atomic mass**

= (atomic no. + no. of neutrons)

= (6 + 6) = (12)

Carbon atom

**Atomic Weight (Mass) of Element:** The atomic mass of an element is how heavy it is. It is made up of protons and neutrons that are in the middle of the element. Some elements have different versions with different amounts of neutrons, but they still have the same amount of protons. The atomic mass is the average weight of all these versions of the element.

**Allotrope:** Allotropes are different forms of an element that look and act different, but are made of the same stuff. Some elements have more than one form. For instance, carbon can be a shiny diamond or a gray pencil lead called graphite.

**Isotope:** Isotopes are different types of atoms that have the same parts, like protons and electrons, but they have a different number of neutrons. For example, the three most stable isotopes of hydrogen: protium (A = 1), deuterium (A = 2), and tritium (A = 3).

**Crystalline Structure of Element:** The crystalline structure of an element is how its atoms, ions, or molecules stick together in a pattern to make a cool crystal shape.

**Ferrous and Non-Ferrous Metals:** When we say ferrous metal, it means that iron is a big part of the metal. But if there's only a little bit of iron in the metal, we call it non-ferrous. The word "ferrous" comes from Latin and means iron, which is why iron's symbol is Fe.

**Ductile Metals:** These are capable of being made into long, thin wire or thread. Copper and Silver are ductile metals.

**Malleable Metals:** These can be hammered or rolled into thin sheets without cracking or breaking. Gold is malleable.

**Ferromagnetic:** Materials that are strongly attracted to a magnet. Such materials can be permanently magnetized. These include the elements iron, nickel and cobalt and their alloys, some alloys of rare-earth metals, and some naturally occurring minerals such as lodestone.

**Magnetostriction—**Ther is the term for a special thing that happens to magnetic materials. When these materials get turned into magnets, they also change their shape or size.

**Paramagnetic:** Slightly attracted to a magnetic field, but do not retain magnetic properties once the field is removed.

**Diamagnetic:** Slightly repelled by a magnetic field, but do not retain magnetic properties once the field is removed.

**Electrical Properties:** Conductor—a thing that lets electricity flow through it. Semi-conductor—a special material (usually silicon) that can conduct electricity, but not as well as metal. Insulator (non-conductor)—a material (usually glass) that stops electricity from flowing.

**Reactive Gas:** These gases are really good at reacting with stuff! They are called "sticky gases" because they can react to things like plastic and wet surfaces when they touch them. These are nitrogen, oxygen, hydrogen, carbon dioxide, fluorine, and chlorine.

**Non-Reactive Gas:** An inert gas is like a super shy gas that doesn't like to hang out with other chemicals. It doesn't make any new friends by reacting with them, so it doesn't form any chemical compounds. We also call these special gases "noble gases."

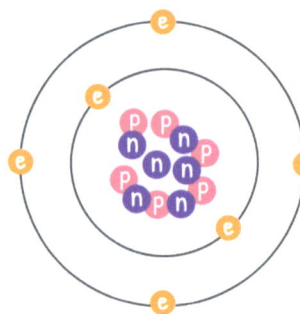

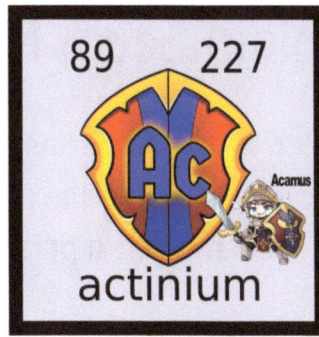

89    227

Ac

Acamus

actinium

# What Makes Actinium Seem Magical?

Imagine a world where elements hold magical powers, transforming the ordinary into the extraordinary. Among these elements, Actinium stands out as something truly special. If we view Actinium through the lens of magic, its abilities may seem almost mystical. Here's a look at how Actinium might appear to have magical powers and why it could be so fascinating.

First, consider the idea that Actinium could amplify and enhance the powers of other elements and individuals. Just like a spell that makes a simple element much stronger, Actinium could be seen as a source of energy that brings out the best in what's around it. For instance, if a magical knight had a tiny piece of Actinium, they might find that their spells become much more powerful. The element could act like a magical booster, helping to elevate their natural abilities and those of other items. Can you imagine casting a fire spell that suddenly bursts into a brilliant display, thanks to the presence of Actinium? This enhancement could make any magic user feel like they have the power of a wizard from a great story.

Now, let's talk about energy manipulation. If Actinium indeed had magical abilities, it could control energy in ways that we can only dream of. Picture this: a magical knight uses Actinium to create brilliant shields of energy that protect them from harm. These protective barriers could turn away negativity, much like a guardian angel of energy, ensuring that the knight remains safe while they work their spells. With Actinium in their possession, anyone could feel secure. This ability not only fascinates the imagination but also adds a practical layer to the idea of magic, turning Actinium into a protector of sorts.

Another enchanting quality of Actinium could be its ability to unlock hidden knowledge and secrets. In stories of magic, there are often ancient scrolls and forgotten truths that very few have the strength to uncover. Imagine if Actinium could reveal these secrets! Those who possess this element might find themselves seeing beyond the physical world. They could glimpse into mystical realms filled with wisdom and power. It's as if Actinium opens a door to a library of the universe's secrets, allowing the user to seek answers that are otherwise hidden. This aspect of Actinium could make it a prized possession for those who thirst for knowledge and adventure.

Actinium might also be seen as a powerful cleanser able to detect and purify negative energies. In our world, sometimes negative feelings or forces can surround us, and these can bring about trouble or unhappiness. If Actinium could sense these harmful energies, it would be like having a magical compass guiding you away from danger. Moreover, its purifying capabilities could transform dark feelings into light and positivity. As such, anyone using Actinium might find that their environment becomes brighter and more supportive, making the element invaluable for personal well-being. In stories filled with magic, having a tool that can remove bad energy would greatly empower the heroes to fight against evil.

In addition to all these powers, Actinium could enhance physical strength, courage, and protection. Imagine holding a small piece of this element and feeling a surge of energy fill your body. Individuals with Actinium might find themselves with enhanced physical abilities, allowing them to perform extraordinary feats. Want to lift something heavy or run faster? The magical might of Actinium could provide that extra boost needed to overcome any obstacle.

Acts of bravery could be another gift from Actinium. When facing challenges, it can be tough to gather the courage to move forward. But with Actinium at your side, you might feel empowered to stand tall and push through even the most intimidating situations. This particular power could inspire hope and determination, characteristics that feel almost magical in themselves. Imagine a hero in a grand tale, standing firm against darkness, and feeling the strength flowing from their Actinium.

Overall, if Actinium had magical powers, its abilities would be connected to strength, protection, and empowerment. It could enhance the existing skills of magical users, create protective barriers, reveal hidden truths, and purify negative energies. For those who possess it, Actinium could be a source of courage and vitality, enabling them to take on formidable challenges. Through this lens of magic, Actinium transforms from a mere element into an enchanting force, sparking imagination and wonder about the magical possibilities that lie within our universe. Just think of all the adventures waiting to unfold with Actinium in the mix!

# Potential Future Uses For Actinium

Scientists and science fiction writers alike have envisioned various intriguing future applications for actinium, a rare and fascinating element of the periodic table. Central to many of these discussions is the isotope actinium-225, which is capturing a lot of attention due to its remarkable potential in the field of cancer treatment. Actinium-225 is a radioactive substance with the unique ability to selectively target cancerous cells and deliver localized radiation therapy, which can effectively kill malignant cells while sparing the surrounding healthy tissue. This precision makes it a promising candidate for addressing certain types of cancers that are notoriously difficult to treat, such as prostate and brain cancers, presenting a revolutionary shift in how we approach oncology.

Beyond its role in cancer therapy, actinium-225, along with other isotopes of actinium, is poised to revolutionize the development of novel medicines. These could serve dual purposes—both in imaging to locate health issues and in treatment to address diseases. For instance, by attaching actinium isotopes to antibodies or other targeting moieties, researchers can create powerful diagnostic tools and therapeutic agents that illuminate or attack specific disease sites in the body. This dual functionality could hold significant implications for early diagnosis and targeted drug delivery systems, enhancing the effectiveness of treatments for various conditions, including not only cancers but also autoimmune diseases and infectious diseases.

Moreover, one of the more unconventional yet promising applications of actinium lies in the creation of small nuclear batteries, often referred to as betavoltaics. These batteries could operate in environments where conventional batteries would struggle or be impractical,

# Potential Futures Uses For Actinium (Continued)

such as in remote locations or in applications requiring long-lasting energy solutions, like pacemakers or space exploration equipment. The longevity and reliability of actinium-powered batteries could provide power for many years without the need for replacement, tackling some of the logistical challenges of energy supply in isolated or critical settings.

The unique properties of actinium also present exciting prospects for scientific research. In fields such as nuclear physics, actinium's isotopes can serve as valuable tools for studying nuclear reactions and understanding fundamental scientific principles. Additionally, its applications in materials science might lead to innovations in creating new materials with tailored properties for various industries, thereby stimulating progress across multiple sectors.

In terms of environmental applications, actinium's radioactive qualities could play a crucial role in addressing the perilous challenge of nuclear waste cleanup. By harnessing its ability to break down harmful radioactive materials, actinium may provide a method for rendering environmental pollutants less hazardous, contributing to more sustainable methods of waste management.

Furthermore, actinium isotopes could be utilized in various industrial applications, including sterilization processes in healthcare settings and quality control in manufacturing. Their potency in targeting and eliminating pathogens makes them ideal candidates for ensuring sterility and safety in products that require rigorous hygiene standards.

In summary, actinium holds remarkable potential that transcends the boundaries of medicine, environmental management, and industrial processes. As research continues to unlock its secrets, we may discover even more innovative ways to leverage this element for the benefit of humanity. Whether addressing cancer treatment, developing advanced diagnostic tools, creating sustainable energy sources, or contributing to environmental cleanup efforts, actinium's multifaceted applications signal a bright future filled with possibilities waiting to be explored. As we deepen our understanding of this element, the transformative impacts of actinium may reshape not only scientific landscapes but also everyday life in profound ways.

# Meet Acamus, The Knight With The Actinium Tipped Sword

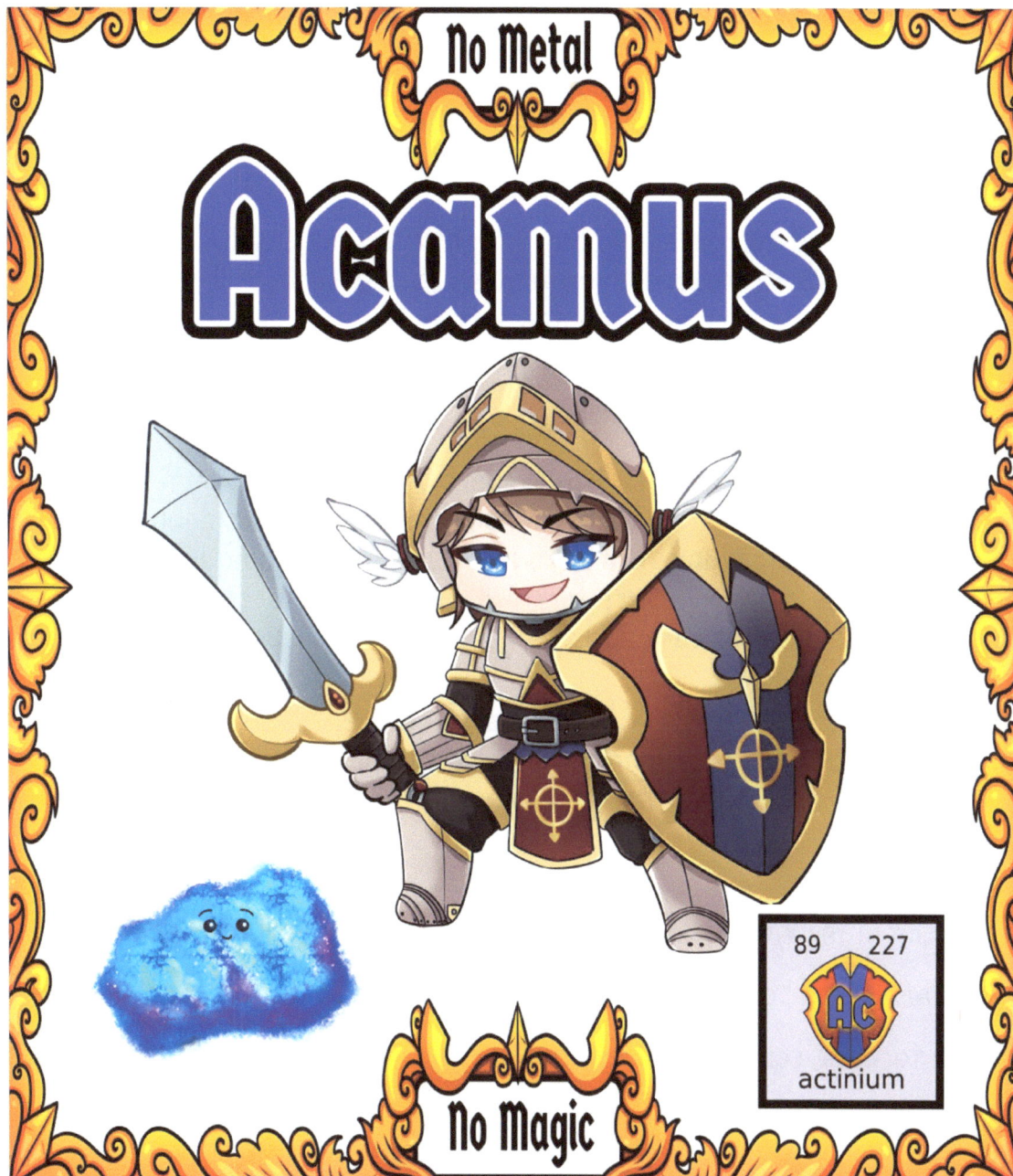

No Metal

Acamus

No Magic

89  227
Ac
actinium

In the land of MarBryn, where the skies glimmered with stars and the air buzzed with magic, there lived a knight named Acamus. Standing small but bold, he was a plucky bundle of courage wrapped in shining armor. His armor glinted under the sun, a sight that inspired hope in the hearts of the people. A gleaming helm adorned with a curved crest framed his determined blue eyes, and a confident, mischievous grin danced on his lips, hinting at the playful spirit tucked beneath the steel exterior.

Acamus gripped his sword, a magnificent weapon forged from a rare metal known as Actinium. The sword sparkled and shimmered, its broad blade gleaming like the morning sun. The guard curled like a crescent moon, and the hilt felt sturdy in his hand. Beside him, a sturdy shield bore a red-and-blue emblem with shiny gold accents, showcasing a bold cross in its center—a symbol of bravery and protection.

Though Acamus was small compared to other knights, he wore lightweight plate armor fitted with leather straps that allowed him to move quickly while still providing protection. What made him stand out even more were the feathered wings tucked at the sides of his helmet, ethereal and whimsical, just like the knight himself. He stood ready, vigilant, and heroic.

One fine morning, the kingdom awoke with whispers of danger. A powerful sorcerer had arisen from the dark corners of the land, his name sending shivers down spines. Magh, the shadowy sorcerer, sought to plunge MarBryn into darkness, tapping into the dark energies that roamed the woods. The people were terrified, and rumors spread that Magh had uncovered a way to harness the dark powers of Actinium for his own wicked benefits.

Hearing the distress of the townsfolk, Acamus straightened his back and tightened his grip on his sword. "Fear not!" he declared, his voice strong yet warm. "I will stop Magh and protect our home!" With that, determination surged through him. It was time for an adventure.

His journey began in the Whispering Woods, where ancient trees whispered secrets to the wind, and shadows lurked behind every corner. Acamus, fueled by the magic of his Actinium sword, pressed forward, sensing the way energy pulsed around him. With each step, he felt the sword's power amplify his own courage, filling him with an irresistible force.

Suddenly, a gust of wind rustled the branches above, revealing, not Magh but one of his minions, Zorath, ahead, cloaked in black robes that swallowed the light. His eyes gleamed, filled with malice. "Ah, the brave knight thinks he can stop me!" he laughed, a chilling sound that echoed through the trees. "With the power of Actinium, I shall help Magh unleash an army of shadows upon your kingdom!"

But Acamus was unafraid. He drew his sword, and a bright light erupted from it, revealing the true magical potential of Actinium. Energy crackled in the air, bolstering his strength and courage. "I won't let you harm my people!" he shouted, the smile never leaving his face even as determination hardened his gaze.

With a swift motion, Acamus surged forward, the wings of his helmet flapping majestically and sending him into the air. The elements of the forest seemed to rally behind him, bolstered by Actinium's magic. As he swung his sword, beams of light shot forward like arrows, striking the shadows that Zorath commanded. The dark creatures let out wails as they returned to the dark corners from whence they came.

But Zorath regained his focus and summoned more shadows. "You think your light can defeat me?" he sneered, raising a hand to conjure a dark force. Acamus felt the darkness pressing in, but he remembered the magic within his sword. With all his might, he focused on the pure energy of Actinium. The air around him shimmered as a protective barrier formed, shielding him from the sorcerer's attack.

Realizing the magic of Actinium could not only defend but purify, Acamus forged ahead, unleashing waves of light infused with positive energy. The shadows recoiled, and Zorath's wicked grin faltered. "What is this power?" he gasped, uncertain of facing such formidable light.

With a final surge of courage, Acamus soared towards the dark sorcerer, his sword glowing like a star. "This is the power of hope and bravery, Zorath! You and Magh will not win!" With one mighty swing, he struck Zorath's spell with the blade of Actinium, and as their energies collided, a brilliant explosion of light erupted. The dark shadows dissolved, fleeting into nothingness as Zorath stumbled back.

Confronted with pure light and strength, Zorath faltered. "No! It cannot end like this!" he shouted, before a flash of light enveloped him, banishing him from the woods. In that instant, the Whispering Woods breathed a sigh of relief as the air cleared, and nature's whispers turned into joyous songs.

Acamus hovered in the air for a moment, letting the exhilaration wash over him. The magical properties of Actinium had not only protected him but had also revealed his true strength as a warrior and a knight. He was victorious, not just because of his sword, but due to the bravery and hope that resided within his heart.

When he descended back to the ground, the townsfolk greeted him with thunderous cheers. The land of MarBryn was safe once more, and Acamus, the small but bold knight, stood tall among them, ready for whatever adventure awaited next.

From that day on, tales of Acamus spread far and wide. The magical powers of Actinium had helped him save MarBryn from one of Magh's minions, but it was ultimately his courage, determination, and friendship with the people that made him a true knight. The world was full of magic, but the greatest magic of all was the strength found in hope and unity. And thus, Acamus, the plucky knight with a heart full of courage, continued to soar through adventures that would one day lead him to the powerfully evil sorcerer, Magh.

# Enjoy This Coloring Page Featuring

# Acamus The Knight With The Actinium Tipped Sword

## Acamus Presents Actinium

Symbol: Gd   Atomic Number: 89   Atomic Mass: 227

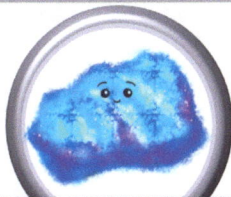

### Actinium Facts

- Discovered in 1899 in Paris
- Silvery white metal
- Radioactivity makes it glow an eerie blue
- Does conduct electricity and heat but radioactive
- Paramagnetic
- Actinide Metal

No Metal

**Acamus**

The Knight With The Actinium Sword

No Magic

Actinium is not found in large quantities in nature but some common ores are Uraninite and Thorite. It is primarily produced through the neutron bombardment of radium in nuclear reactors. The radium absorbs a neutron and undergoes beta decay to form actinium.

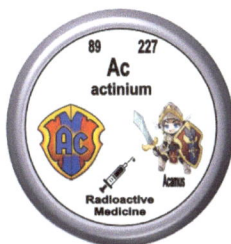

89   227

Ac

actinium

Radioactive Medicine

### Acamus' Magical Abilities

Acamus uses his sword to control the powers of Actinium. It can create icy winds to stop bad guys or shoot lightning to hit targets perfectly. He can also use it to make shields of blue light to protect himself and his friends.

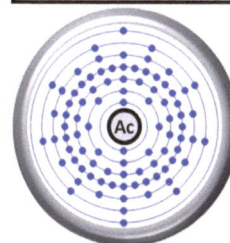

Ac

### Atomic Structure

### Uses For Actinium

There are very few current uses for Actinium but as the first element in the actinide series, it shares many chemical properties with the lanthanides, particularly lanthanum. Forward thinkers propose some interesting commercial potentials for its use in the future.

Futuristic cars might use Actinium isotopes in massive HEPA rated automotive air filters.

Actinium isotopes could be used in autonomous military robots for long-range missions.

Actinium-225 has been used to target cancerous bone cells for treatment.

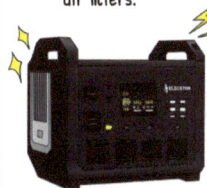

Actinium-powered energy sources could be used for off-grid applications.

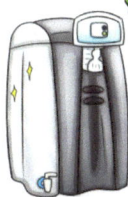

Actinium isotopes might be used to break down contaminants in water supplies.

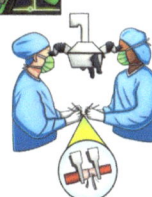

Actinium isotopes are actively being researched for minimally invasive radiation treatments.

### Did You Know?

- Actinium is the first element in the actinide series. The name actinide is derived from the word Actinium.
- The word "actinium" from the Greek "aktis," which means "sunbeam" — an allusion to its high radioactivity.
- Actinium-227 emits beta particles with so much energy that it lights up the surrounding environment with a blue glow. In a sense, due to its high levels of radioactivity, it is vibrating faster than the speed of light and the blue glow is like a sonic boom.
- The only naturally occurring isotope of actinium is 227Ac. Thirty-six radioisotopes of actinium have been identified, all with half-lives ranging from 69 nano seconds at the shortest (for 217Ac) to 21.77 years at the longest (227 Ac).
- Actinium-225, is being used a new type of cancer treatment called targeted alpha therapy. As the atom decays, it releases particles that destroy the nearby cancer cells, but don't travel far enough to damage the rest of the body.

# MEET THE ACTINIDE KNIGHTS

| 89  227 |
|---|
| **Ac** |
| actinium |

| 95  243 |
|---|
| **Am** |
| americium |

| 97  247 |
|---|
| **Bk** |
| berkelium |

| 98  251 |
|---|
| **Cf** |
| californium |

Acamus—Actinium
(Not a Transuranic Element)

Amerine—Americium
(Transuranic Element)

Beremere—Berkelium
(Transuranic Element)

Calastian—Californium
(Transuranic Element)

| 96  247 |
|---|
| **Cm** |
| curium |

| 99  252 |
|---|
| **Es** |
| einsteinium |

| 100  257 |
|---|
| **Fm** |
| fermium |

Curran—Curium
(Transuranic Element)

Elizama—Einsteinium
(Transuranic Element)

Ferley—Fermium
(Transuranic Element)

| 101  258 |
|---|
| **Md** |
| mendelevium |

| 102  259 |
|---|
| **No** |
| nobelium |

| 103  266 |
|---|
| **Lr** |
| lawrencium |

| 93  237 |
|---|
| **Np** |
| neptunium |

Larelis—Lawrencium
(Transuranic Element)

Menesant—Mendelevium
(Transuranic Element)

Nepthas—Neptunium
(Transuranic Element)

Tellan-Tellurium
(Transuranic Element)

| 94  244 |
|---|
| **Pu** |
| plutonium |

| 91  231.0 |
|---|
| **Pa** |
| protactinium |

| 90  232.0 |
|---|
| **Th** |
| thorium |

Puchan—Plutonium
(Transuranic Element)

Protik—Proactinium
(Not a Transuranic Element)

Thordis—Thorium
(Not a Transuranic Element)

# THEY WIELD SWORDS, LANCES AND MACES POWERED BY ELEMENTS FROM THE PERIODIC TABLE

# Create Your Own Magical Actinide Knight Elemental

## Acamus — The Knight With The Actinium Sword

Symbol: Ac  Atomic Number: 89  Atomic Mass: 227

Magical Elemental Symbol

89 227
**Ac**
actinium

Actinium is a Actinide Metal

Found in some common ores like Uraninite and Thorite but mainly produced in nuclear reactors

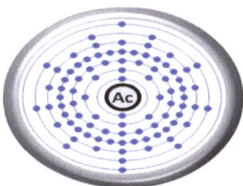

89 227
Ac actinium
Radioactive Medicine

Actinium Periodic Symbol

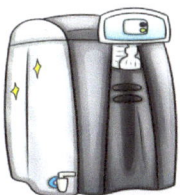

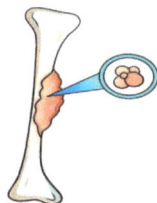

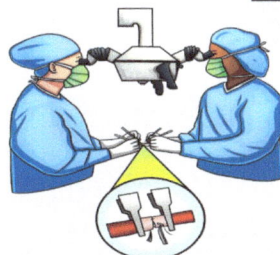

Acamus's Magical Abilities

Acamus uses his sword to create icy winds to stop bad guys or to shoot lightning that hits targets perfectly.

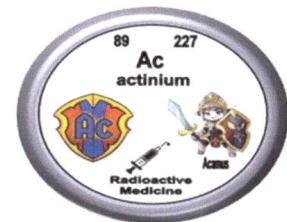

Atomic Structure

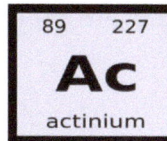

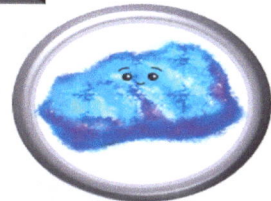

# Magical Elements of The Periodic Table

## Students may either use a program like power point to cut and paste clip art into a Magical Knight Elemental Blank or, if they wish, they may draw everything themselves.

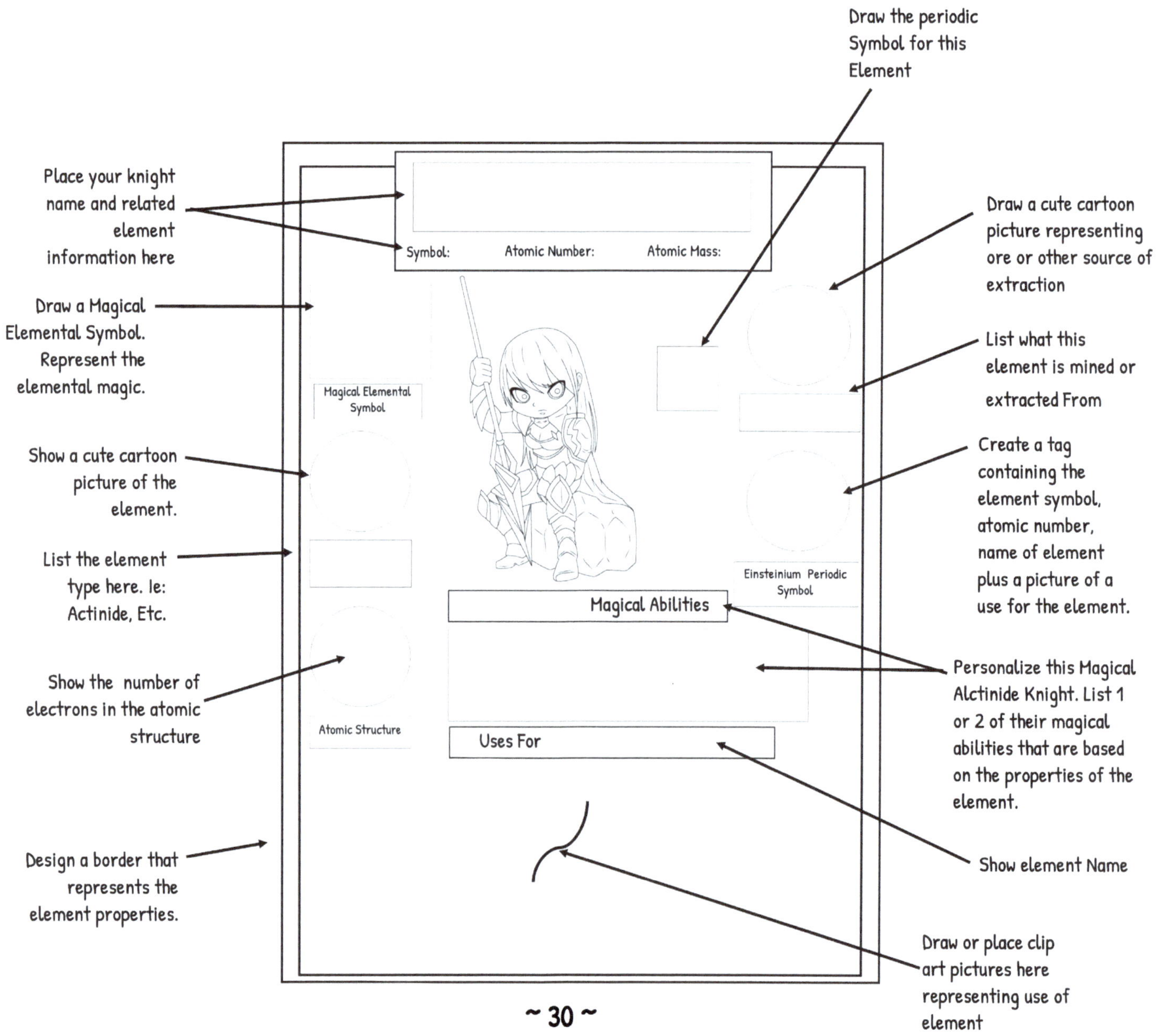

Draw the periodic Symbol for this Element

Place your knight name and related element information here

Symbol:          Atomic Number:          Atomic Mass:

Draw a cute cartoon picture representing ore or other source of extraction

Draw a Magical Elemental Symbol. Represent the elemental magic.

Magical Elemental Symbol

List what this element is mined or extracted From

Show a cute cartoon picture of the element.

Create a tag containing the element symbol, atomic number, name of element plus a picture of a use for the element.

List the element type here. Ie: Actinide, Etc.

Einsteinium Periodic Symbol

Magical Abilities

Show the number of electrons in the atomic structure

Atomic Structure

Personalize this Magical Actinide Knight. List 1 or 2 of their magical abilities that are based on the properties of the element.

Uses For

Design a border that represents the element properties.

Show element Name

Draw or place clip art pictures here representing use of element

Symbol:                Atomic Number:                Atomic Mass:

Magical Elemental
Symbol

Actinium  Periodic
Symbol

Magical Abilities

Atomic Structure

Uses For

# Magical Knight Elemental Research Sheet

Before starting your Magical Knight Elemental graphics page, do some research on your chosen element.

| | |
|---|---|
| Name of Magical Knight: | |
| Knight's Magic Power Based on the Element's Properties: | |
| Magical Elemental Symbol: | |
| Element Name: | |
| Element Symbol: | |
| Atomic Number: | |
| Atomic Mass: | |
| What year and where was this Element discovered? | |
| Who discovered this Element? | |
| Element Group: | |
| Element Period: | |
| Element Family Name: | |
| State of Element At Room Temperature: | |
| What is Element Mined or Extracted From? | |
| Is Element Magnetic? | |
| Does Element Conduct Electricity? | |
| Where is the Element commonly found in Nature? | |
| What is 1 alloy of the Element? How used? | |
| What is 1 compound of the Element? How used? | |
| Name the most common use for this Element: | |
| Name a little known use for this Element: | |
| Name one more use for this Element: | |
| Interesting and Fun Facts: | |

# Magical Unicorn Elemental Research Sheet

Before starting your Magical Unicorn Elemental graphics page, do some research on your chosen element.

| | |
|---|---|
| Name of Magical Unicorn: | Ghel The Gold Horn Unicorn |
| Unicorn's Magic Power Based on the Element's Properties: | Ghel can see past, present and future. She is empathic and can sympathize with the feelings of other. They say she has a heart of gold. |
| Magical Herd Crest Symbol: | An open heart with a Celtic Trinity Knot. |
| Element Name: | Gold |
| Element Symbol: | Au— Comes from Aurum which is the Latin word for Gold. |
| Atomic Number: | 79 |
| Atomic Mass: | 196.97 |
| What year and where was this Element discovered? | Around 4,600 BCE in Bulgaria |
| Who discovered this Element? | Unknown |
| Element Group: | 11 on Periodic TAble |
| Element Period: | 6 on Periodic TAble |
| Element Family Name: | Gold is a Noble Transition Metal |
| State of Element At Room Temperature: | Solid |
| What is Element Mined or Extracted From? | Quartz Veins. It is also found in gravel in streams. |
| Is Element Magnetic? | It is Diamagnetic. It's only weakly magnetized when placed in a magnetic field. |
| Does Element Conduct Electricity? | Gold is a great electrical conductor used in printed circuitry of computers. |
| Where is the Element commonly found in Nature? | One of the largest deposits is found in the United States in Arkansas. |
| What is 1 alloy of the Element? How used? | White gold is an alloy of gold, palladium, nickel and zinc. |
| What is 1 compound of the Element? How used? | Gold Phosphide is a semiconductor used in high power, high frequency applications and in laser diodes. |
| Name the most common use for this Element: | Jewelry |
| Name a little known use for this Element: | Acupuncture needles |
| Name one more use for this Element: | Gold is used in airbags in cars. |
| Interesting and Fun Facts: | Gold was used in ancient Egypt to fill decayed teeth. Gold thread is incorporated in astronaut spacesuits to protect them from the heat of the sun. |

Write a paragraph below to describe your magical knight elemental. Based on the information obtained from research of your chosen element, how did you determine your knight's name? What are your knight's magic powers? What are their likes/dislikes, strengths/weaknesses, personality traits? What color are your knight's armor and clothing and why did you pick that/those colors? What is your knight's Magical Elemental Symbol?

# Magical Unicorn Elemental

# Sample Description

## Ghel The Gold-Horned Unicorn

This magical unicorn has a golden horn and hooves that glow like the sun. Her hide is honey-gold and her flowing mane and tail are golden-blonde.

Ghel is a member of the Metal Horn Unicorn Tribe from Unimaise. Gold is linked to the heart chakra because it holds a warm energy that brings soothing vibrations to the body to aid in the healing process. Ghel's Magical Herd Crest symbol is an open heart with a Celtic Trinity knot which indicates that she can see past, present and future. She is empathic and can sympathize with the feelings of others.

Being empathic does not make Ghel a weakling. She is brave with strong opinions and is a true champion to those she loves. She is known as the unicorn with the "heart of gold". When she places her horn on the heart of another, she senses their future.

The name Ghel is an Indo-European word which means yellow. The word "gold" most likely has its origins in the word "Ghel".

Gold is known to possess spiritual powers that bring happiness, peace, stability and luck to those who wear it. Scientists say that all the gold in the world comes from the collision of neutron stars.

Though most other magical unicorn elementals get along well with the gold horn unicorn herd; Ghel, and others like her, must be very careful around the Quick Silver Herd - as gold dissolves in mercury.

# Do Your Middle Graders Want To Know More From The Magical Elementals About The Periodic Table?

Get the accompanying books in print at all online book stores. Get the books and accompanying activities at MagicalPTElements.

## Available Now

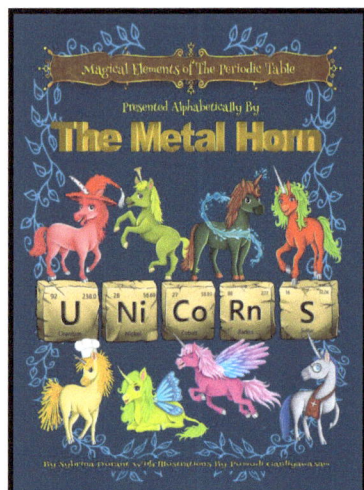

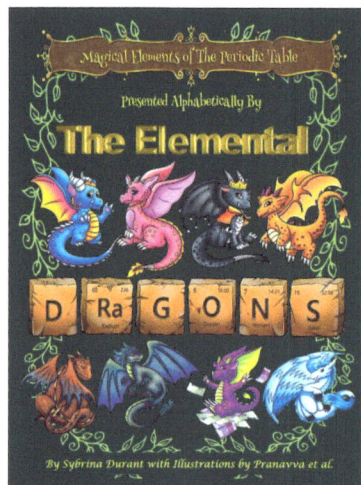

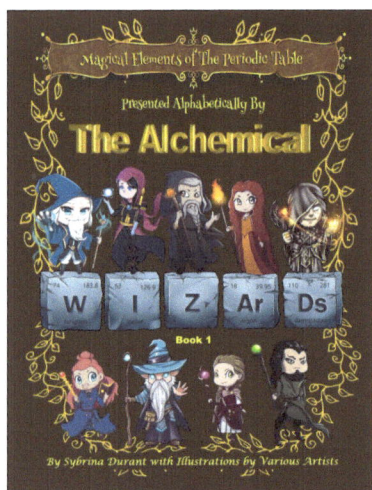

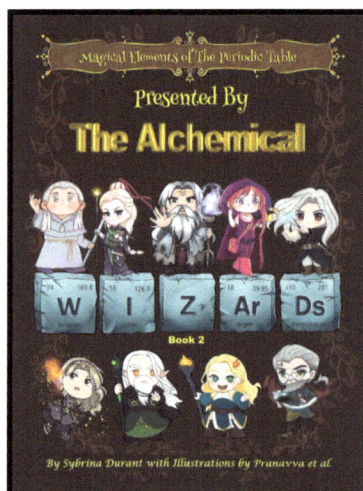

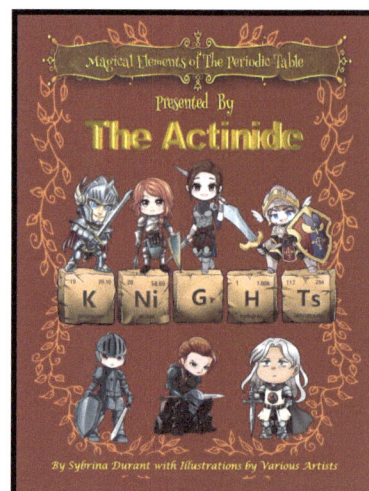

## Coming Soon

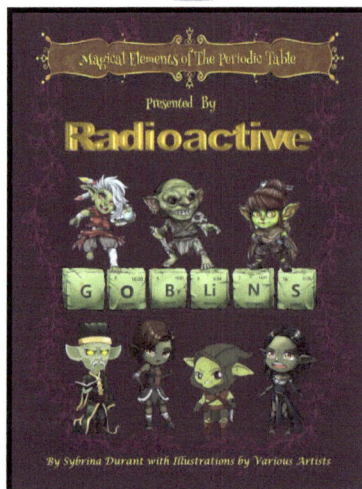

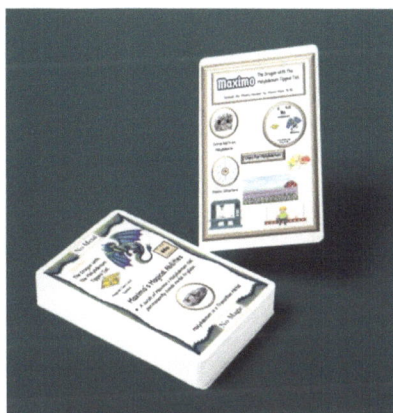

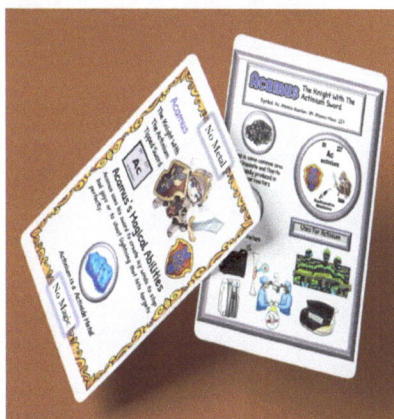

Learn More About all of the Periodic Table Elementals.
Get all of the "No Metal No Magic" Books Featuring
Individual Elements at MagicalPTElements.com

No Metal No Magic
Book 1

Aluminum Presented by Alumna
from The Magical Elements of the Periodic
Table Series

13   26.98
**Al**
aluminium

Alumna

Aluminum

By Sybrina Durant with Illustrations by Pumudi Gardyawasam

No Metal No Magic
Book 26

Antz, From The Magical Elements of the
Periodic Table Series,
Presents Antimony

Antz

51   121.8
**Sb**
antimony

Antimony

By Sybrina Durant with Illustrations by Pranavva et al.

No Metal No Magic
Book 2

Cuprum, from The Magical Elements of the
Periodic Table Series
Presents Copper

Cuprum

29   63.55
**Cu**
copper

Copper

By Sybrina Durant with Illustrations by Pumudi Gardyawasam

No Metal No Magic
Element 2

Helium, Presented By Hetha
From The Magical Elements of the Periodic
Table Series

Hetha

2   4.003
**He**
helium

Helium

By Sybrina Durant with Illustrations by Pranavva et al.

# Also Available From Sybrina Publishing
# Magical Elemental-Themed Periodic Table

Magical elementals from the Magical Elements of the Periodic Table books present all of the elements of the periodic table in fantastical and real life terms.

In the books, each elemental character has magical powers based on the properties of the elements that come from the land, air and water. They are the perfect group to introduce you to metals, metalloids, non-metals, halogens, noble gases and much more.

Unicorns, dragons, alchemists, knights, and goblins will show you how people of this world always have and always will depend upon the elements that our earth provides for all of our needs.

Use this Periodic Table as you would any other to spark an interest in the magical and real world properties of all the elements known today. You may be surprised at how prominently they feature in our every day lives.

Magical Elements of The Periodic Table

## No Metal
### No Magic
Actinium To Zirconium

Remember, "No Metal— No Magic." . . .And no technology.

It's Techno-Magical

SUPER HEAVY METALS—RADIOACTIVE

RARE EARTH LANTHANIDE METALS

ACTINIDE METALS

## LEGEND
- Alkali Metals
- Alkali Earth Metals
- Transition Metals
- Post-Transition (or Other Metals)
- Metalloids
- Non-Metals
- Halogens
- Noble Gases
- Rare Earth Lanthanide Metals
- Actinide Metals
- Super Heavy—Radioactive

Alloys are created when 2 or more metals are combined. Compounds are created when 2 or more non-metals are combined.

### EXAMPLE OF A COMPOUND
Quincy

Quick Lime = Ca (Verty, Teeth) + O (Ozzy, Air)
Used for Concrete

Both Carbon and Oxygen are reactive nonmetals.

White Wing

### EXAMPLE OF AN ALLOY
White Gold

White: Ni (nickel), Zn (zinc)
Au (gold), Cu (copper), Ag (silver)
Gold

Used for jewelry, dental amalgams plus connectors, and switch and relay contacts for electronics.

Includes 58.5 % gold, 22% copper, 8% zinc, 7% nickel, 4.5% silver and possibly other elements.

Sybrina.com

Get These Fun Elemental Periodic Table Activities at

MagicalPTElements.

Unicorn Periodic Table
Bingo—Comes with 32
unique Bingo cards.
Magical Elementals
Bingo comes with 36.

Magical Elemental Game
Cards—Makes great prizes.
Fun to trade, too.

plus

1
2
3

Unicorn Horn

Alphabet
CLIP ART FOR YOUR GRAPHICS

A
B
C

Also browse activities at
**https://www.magicalPTelements.com**
for all kinds of printable downloads to make learning fun.

# Printable Magical Elemental Activity Downloads

## Fun Way For Students To Learn The Elements Of The Periodic Table

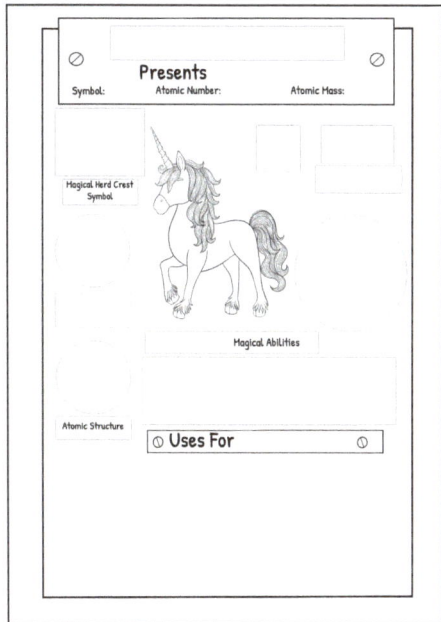

**Blank Unicorn Element Card**

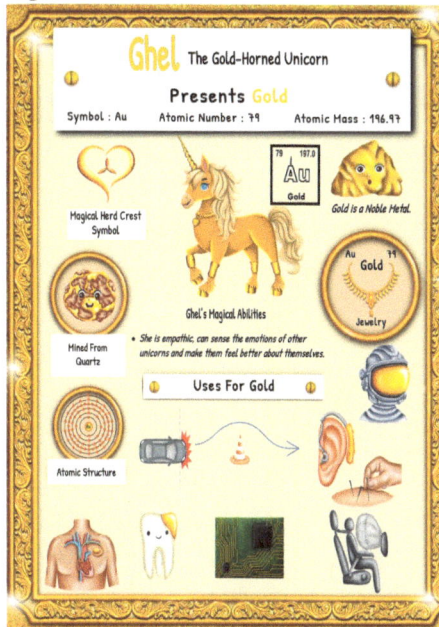

**Sample Unicorn Element Card**

**Blank Research Sheet**

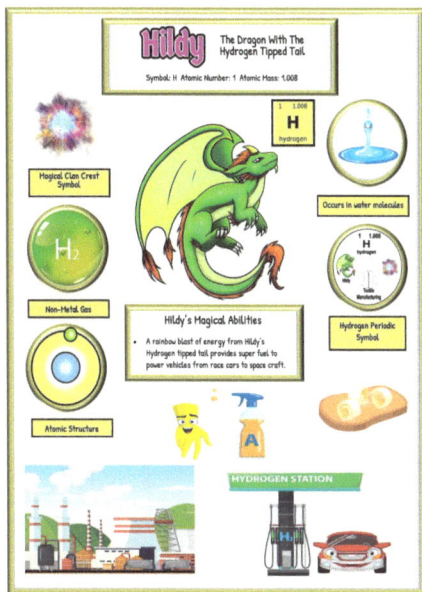

**Sample Dragon Element Card**

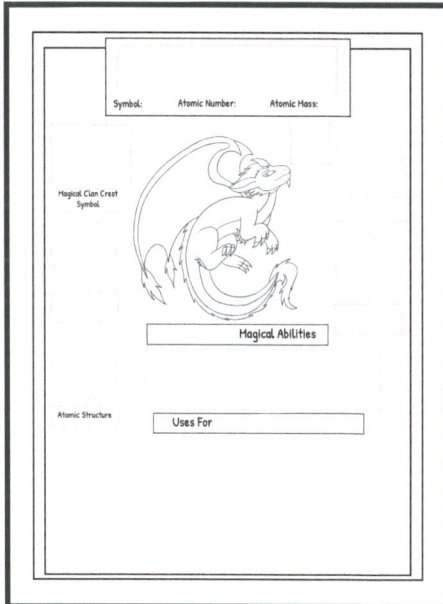

**Blank Dragon Element Card**

**Blank Research Sheet**

Using the sample Magical Elemental cards provided, have students select an element from the Periodic Table and a Magical Elemental Card Blank to create their own Magical Elemental Card. The blank and sample cards do not have to match.

You will receive a pdf containing either 26 unicorn or 26 dragon sample cards and blanks to be printed on 8 1/2 x 11 sized paper or card stock. The pdf also contains a Magical Elemental Research Sheet for the students to work on before creating their unique Periodic Table Elemental. They will also write a short paragraph describing their Unicorn or Dragon Elemental from that research.

Get These Fun Elemental Periodic Table Activity Sheets at **MagicalPTElements.com**

Don't Forget To Get A Tee Shirt

Featuring Your Favorite of the

118 Elements from the Periodic Table

Available in adult and kid sizes in many colors.

https://amzn.to/47NVZWN

This is Acamus' Tee Shirt Graphic.

# Acamus

## The Knight With The Actinium Sword

Symbol: Ac  Atomic Number: 89  Atomic Mass: 227

## Dear Reader

I hope "No Metal No Magic Element 89 — Actinium, Presented By Acamus, from The Magical Elements of the Periodic Table Book Series" with illustrations by Pranavva et al, has helped you learn some fun and interesting things about the magic of the element, Actinium.

This is one of what will eventually be 118 books featuring periodic table elements presented by unicorns, dragons, wizards, knights and goblins. Keep checking regularly. Every one of the elements are amazing and very necessary to our everyday lives. All of the elements are

## Techno-magical.

A lot of research went into every page of this book as well as the Magical Elements of the Periodic Table Books. There are just too many references to publish in this book but you can read and research them all at MagicalPTElements.com/MAUPT or /MDAPT or /MW1PT or / MW2PT or /MAKPT. There, you can also access book related activity sheets and games to help make the learning process more fun.

Get ready made trading cards, lapel pins, tee shirts and more based on this book from Sybrina Publishing's No Metal No Magic Collection at Zazzle - **http://bit.ly/3km64Wg**

Would you like a 24" x 36" poster of the Elemental-Themed Periodic Table in this book? The best place to get it is at **https://bit.ly/49QMxBT** They have the sharpest images of any other poster printer around.

The Magical Elements of the Periodic Table books came into existence because of my Blue Unicorn—Journey To Osm books. If it weren't for their magical powers, based on the properties of the metals of their horns and hooves, I would have never come up with the idea to relate magical creatures to the periodic table. There's a metal horn unicorn story for every age group and they are all available at MagicalPTElements.com

If you enjoyed this book
please leave a nice review
at your favorite online book site.

# No Metal No Magic

## Song Lyrics

No metal, no Magic

No metal, no Magic

I can think of nothing more tragic

Than to have no metal or no magic

Metal makes everything magical.

Just ask a unicorn. . .

Preferably, one with a metal horn.

They'd say No metal, No magic.

Metal makes everything techno magical.

No metal, No magic

for two-leggers or unicorns.

No metal, No magic

Metal makes everything techno magical.

No metal, No magic

It's techno magical.

No metal, No magic

It might be very hard to believe but with

No metal, No magic

There'd be no technology.

No metal, No magic

Listen to this song at https://youtu.be/tcB8KDWAd8w

Watch the book trailer at https://youtu.be/NIX9fE7GJRI

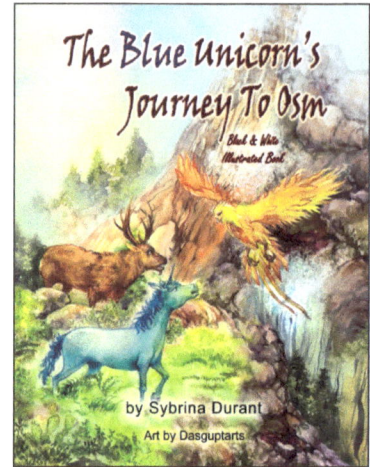

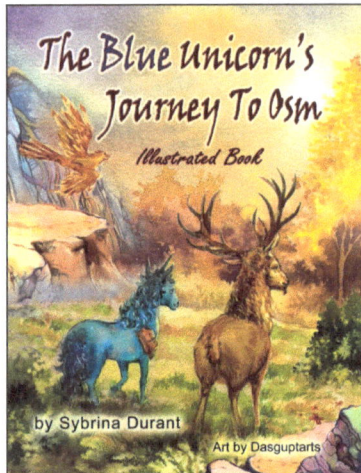

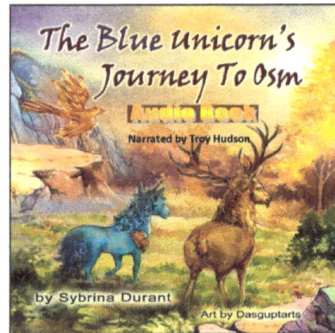

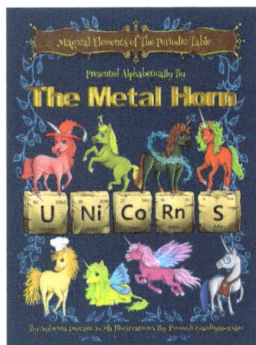

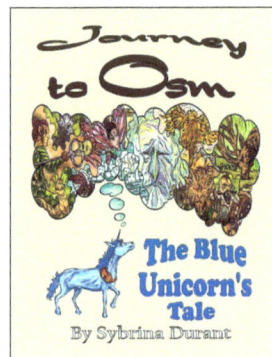

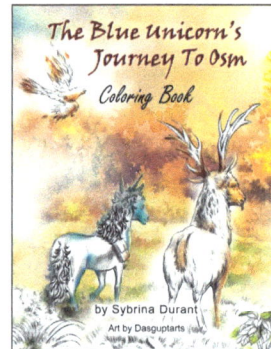

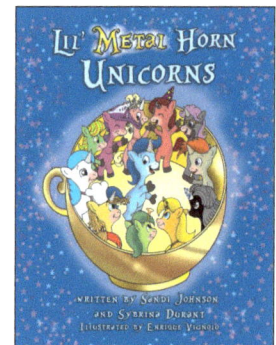

# Journey To Osm—The Blue Unicorn's Tale

Back then, most places throughout MarBryn had wizards and sorcerers of some ability, or another. Most were trained in the ways of magical arts by the unicorns as part of their outreach program. Some two-leggers developed practical magical skills like making delicious feasts of tasty food appear out of thin air or purifying murky water around the land.

Others went through more extensive training to learn battle magic—like shooting powerful streams of energy from their swords.

Some were taught the art of holding the glow of the sun in magical globes, bringing light into the dark of night. These magical lights warded off the evil beings that were new and frightening products of dark magic.

With the rise of the sorcerer Magh, magical defense arts had become more important. The highest level of magical training involved sensing when others were in danger and learning to see into the future. Very few two-leggers ever reached that level because magic wasn't inherent in them the way it was for the unicorns. The metal of their horns and hooves were part of them as well as the very makeup of their blood, but two-leggers relied on learned magic via potions, charms, and incantations that required help from ingredients and forces more mystical in nature than any two-legger was ever born to be. Of course, controlling magic and projecting your intentions went far beyond merely following a recipe of sorts. It took being in touch with nature and the various elements to get the response a wizard desired. Much trial and error went into it, as well as faith and trust and the motives of the spell caster.

Magic was and is a practice that is never quite perfected even for the unicorns who must continue to hone and learn how to harness their powers. On rare occasion a wizard and unicorn had formed enough trust and a steadfast bond that prompted the unicorn to gift the wizard with a wand or staff embedded with the smallest sliver of metal from one of their hooves. This was rare but had happened and of course so had the desire for more power. Magh wasn't the first sorcerer with lust for more and throughout history there had been a handful of heinous acts against unicorns from those seeking their magic. Prior to Magh those wizards had failed to circumvent the protections nature had infused unicorn magic with so even after harvesting metal from their horns or hooves these sorcerers had gone mad trying to bend the will of nature and actually use their ill-gotten gains.

Magh, too, had gone mad or perhaps he'd already been so, but somehow he'd managed to harness the metal he harvested from his victims and continued to grow stronger rather than completely lose his mind like the others. How exactly remained a mystery to the unicorns and everyone else.

The wizards who'd honed their craft with the help and blessing of the unicorns tried to solve the riddle and even the best of them failed to uncover that secret. Some of MarBryn's natives took to magic naturally, while others struggled with the concept but no matter how adept they were. All of them fought valiantly against Magh's magic because with even a shred of knowledge they understood how the power shift would ultimately play out. They fought to the end but, in the end, only one sorcerer remained in MarBryn and now, Magh was in total control. But still, he was not satisfied. He wanted to control every living creature in the land.

www.ingramcontent.com/pod-product-compliance
Lightning Source LLC
Chambersburg PA
CBHW061142030426
42335CB00002B/70